Papa's Beach

Dear Leighton,
I hope you can go to the beach to find treasures too!

Written by:

Lynn Steeves

Illustrations created by Erin Steeves and E. Jackie Brown.

AuthorHouse™
1663 Liberty Drive
Bloomington, IN 47403
www.authorhouse.com
Phone: 1-800-839-8640

© 2011 Lynn Steeves. All Rights Reserved.

No part of this book may be reproduced, stored in a retrieval system,
or transmitted by any means without the written permission of the author.

First published by AuthorHouse 09/09/2011

ISBN: 978-1-4634-0500-7 (sc)

Library of Congress Control Number: 2011916285

Printed in the United States of America

This book is printed on acid-free paper.

Because of the dynamic nature of the Internet, any web addresses or links contained in this book may have changed since publication and may no longer be valid. The views expressed in this work are solely those of the author and do not necessarily reflect the views of the publisher, and the publisher hereby disclaims any responsibility for them.

Dedication

For Mom and Dad because they taught us how to work hard and play hard. Thanks for all the time you spent at various beaches with our family. And to little Miss Macey, your gentle and quiet nature just melts my heart!

Come on Macey, one more big bite to finish your breakfast. It's time to leave for Papa's beach! 1.

Make sure you have both your pink shoes! 1, 2.

Is everyone here? Mommy, Daddy and Macey. Is that three? 1, 2, 3.

Can you count the big mountains? Brrr, they still
have snow on top! 1, 2, 3, 4.

We are finally at the ferry. Let's go on deck where we can see the Pacific Ocean. How many dolphins are playing in the water? 1, 2, 3, 4, 5.

Papa! We are here! And so are Grampa and GG! Now how many people are there? 1, 2, 3, 4, 5, 6.

Let's go to Papa's beach to find some colourful starfish. Did you find all seven? 1, 2, 3, 4, 5, 6, 7.

Look at all the crabs! They sure are good at hiding. Be careful, sometimes they pinch! 1, 2, 3, 4, 5,6,7, 8.

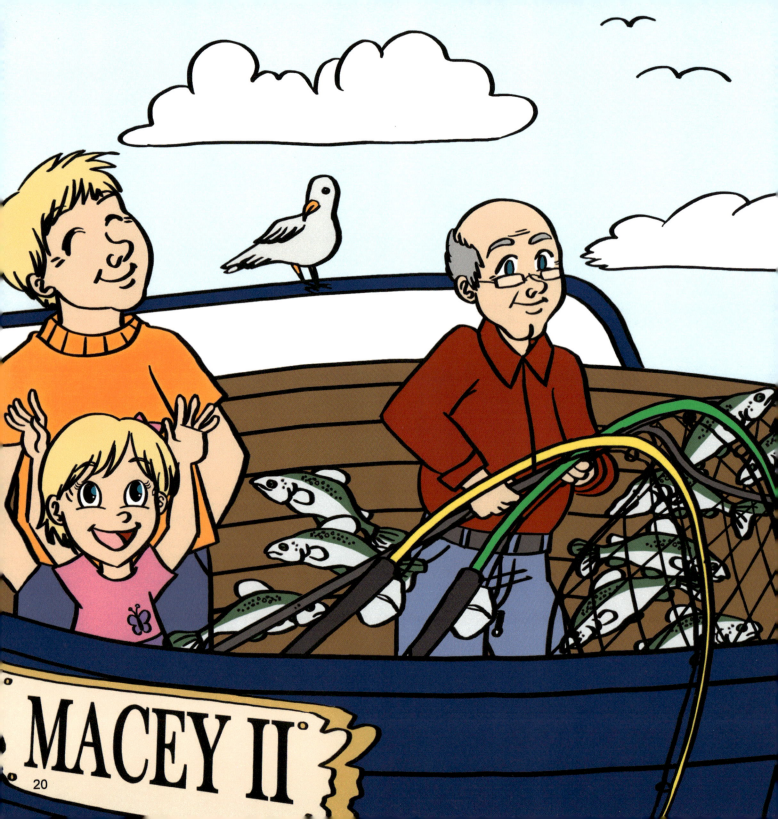

We get to go fishing in Grampa's new boat! I wonder how many yummy salmon we will catch. 1, 2, 3, 4, 5, 6, 7, 8, 9.

Shhh, it is time for bed. Can you count all of the shining stars before falling asleep? 1, 2, 3, 4, 5, 6, 7, 8, 9, 10. Good night Miss Macey. Sweet dreams!

CPSIA information can be obtained
at www.ICGtesting.com
Printed in the USA
256504LV00005B